Microwave Oven Repair

Made Easy

Copyright © 2020

Humphrey Kimathi

All rights reserved.

Disclaimer of liability

The reader of this book is expressly warned to consider and adopt all safety precaution that might be indicated by the activities herein and to avoid all potential hazards.

The author particularly disclaims any liability, loss or risk taken by individuals who directly or indirectly act on the information contained herein.

The author believes that the information presented here is sound, but readers cannot hold him responsible for either the actions they take or the result of those actions.

Table of contents

Safety working on the microwave	3
Basic tools required in microwave repair	6
Components found in microwave	10
Micro switch	10
High voltage circuit	14
High voltage transformer	15
Testing HV transformers	18
High voltage capacitor	20
Magnetron	21
High voltage diode	23
Fuse	26
Thermal cut-out	27
Turntable motor	28
Fan motor	29
Roller guide	30
No heating problems	31
Microwave leakage test	33
Measurement of microwave power	34
Other books by this Author	35

Safety working on the microwave

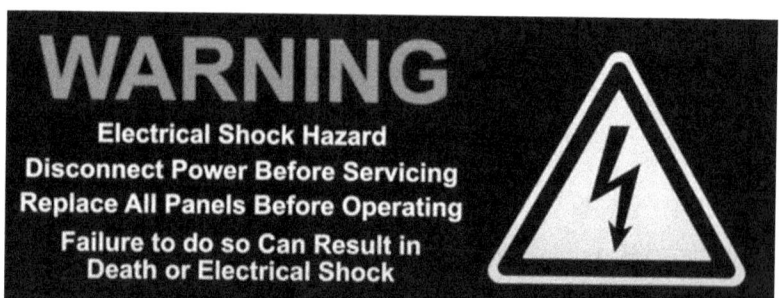

Microwave ovens contain circuitry capable of producing extremely high voltage and current. Contact with the following parts will result in electrocution.

1. *High voltage capacitor*
2. *High Voltage transformer*
3. *Magnetron*
4. *High voltage rectifier assembly*
5. *High voltage wires*

When working on microwave oven always REMEMBER 3D

1) Disconnect the supply
2) Door opened
3) Discharge high voltage capacitor

It is recommended that wherever possible faultfinding is carried out with the supply disconnected and remove the power plug from the outlet. It may in, some cases, be necessary to connect the supply after the outer case has been removed, in this event carry out 3D and then disconnect the leads to the primary of the High voltage transformer. Ensure that these leads remain isolated from other components and the oven chassis by using insulation tape.

- ✓ DO NOT operate on a 2-wire extension cord during repair and use.

- ✓ NEVER TOUCH any oven components or wiring during operation.

- ✓ BEFORE TOUCHING any parts of the oven, always remove the power plug from the outlet.

- ✓ Wait for about 60 seconds after the oven stops, an electric charge remains in the high voltage capacitor. When replacing or checking, you must discharge the high voltage capacitor by shorting across the two terminals with an insulated screwdriver.

- ✓ NEVER operate the oven with no load. Microwave ovens should not be run empty. To test for the presence of microwave energy within a cavity, place a cup of cold water on the oven turntable, close the door and set the power to HIGH and set the microwave timer for two (2) minutes. When the two minutes has elapsed (timer at zero) carefully check that the water is now hot.

- ✓ NEVER injure the door seal and front plate of the oven cavity.

- ✓ NEVER put iron tools on the magnetron.

- ✓ NEVER put anything into the latch hole and the interlock switches area.

- ✓ All input and output microwave connections, waveguide, flange and gasket must be secure never operate the device without a microwave energy absorbing load attached.
- ✓ Never investigate an open waveguide or antenna while the device is energized.

- ✓ Proper operation of the microwave oven requires that the magnetron be assembled to the waveguide and cavity. Never operate the magnetron unless it is professionally installed.

- ✓ Be sure that the magnetron gasket is properly installed around the dome of the tube whenever installing the magnetron.

Basic tools required.

Microwave repair does not require special tools and if you have ordinary tools used to repair television, you are good to go.

Tools come in different qualities. If you cannot afford high quality tools, then you can economise and buy less expensive ones. However, if your budget allows it, purchase the better-quality tools.

Diagonal Cutter

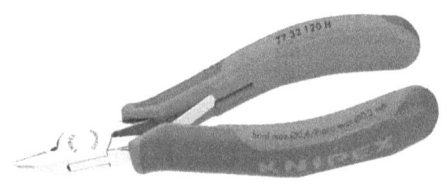

Long Nose Pliers

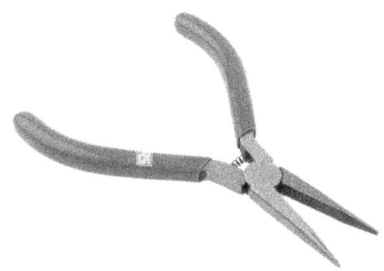

De-soldering

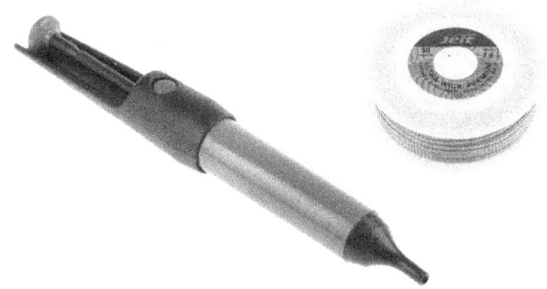

Soldering Station

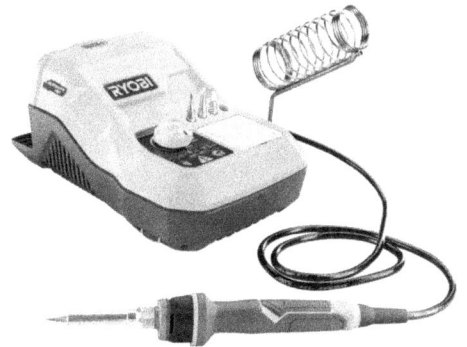

Solder

Screwdriver set

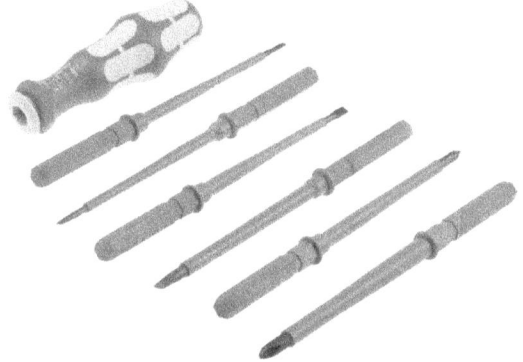

Analogue Multimeter

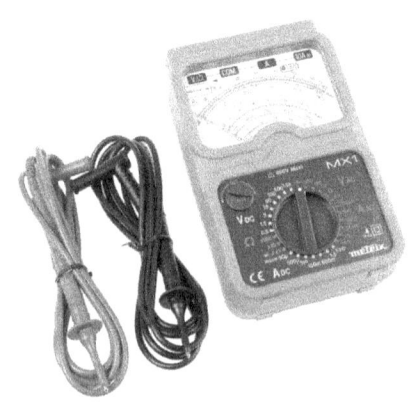

Digital Multimeter

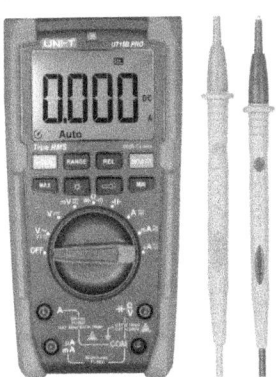

Component and ESR tester

Microwave leakage tester

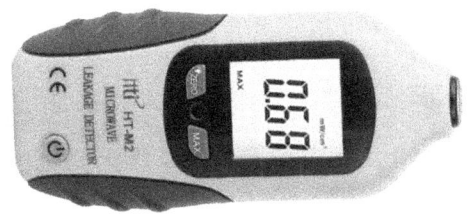

Components found in microwave

First things first.

Once you open a microwave oven make it a practice to take pictures of internal connectors so that when re-assembling you will not have to struggle figuring out which port was for what, this is especially important with the switches, any wrong connection to switches can be devastating.

Components found in microwave, identification, purpose and testing if good or bad.

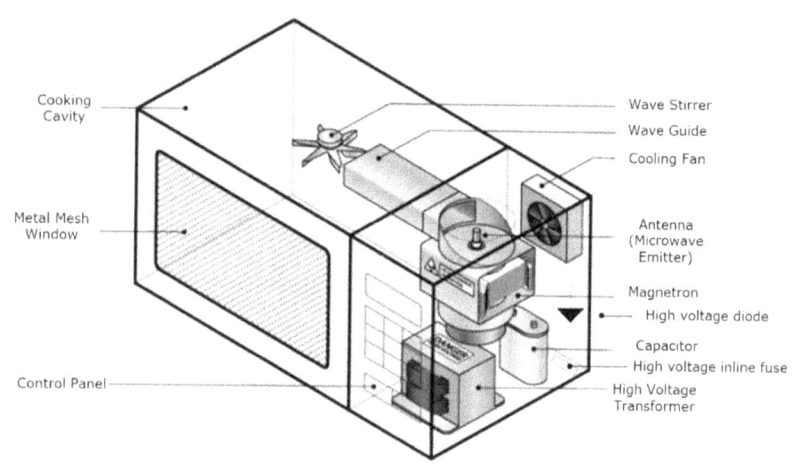

Basic main parts of a Microwave Oven

Interlocks and switches not shown.

Micro switch

A micro switch is an electric switch that is activated with very little physical force.

Microwave oven repair - made easy

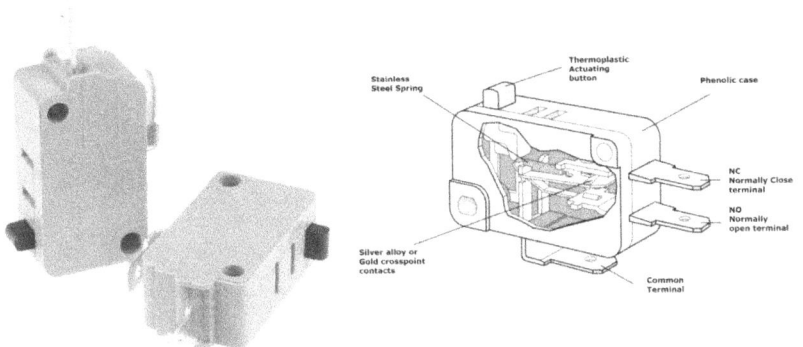

In its simple form, a micro switch has 3 terminals:

Common (COM)
Normally Open (NO) and;
Normally closed (NC).

Common (COM) - Always has power (electricity) so don't confuse with ground, we don't have ground on switches.

Normally closed (NC) - When the micro switch is on its resting position NC is connected to the common.

Normally open (NO) - When the micro switch is pressed, the common connects to the NO (normally open)

When removing a micro switch from the microwave it is important to take a photo of the connecting wires just in case you may forget because inserting the wire to the wrong connector can be devastating.

A microwave switch is used to supply power to the various circuits(components) of the microwave when the door is closed and cut off supply when the door is open. Make sure the microwave does not operate if the door is open which could be extremely dangerous to the user due to risk of exposure of microwave energy. That also explains why you should never connect a jumper wire across the switch if you find it faulty.

There are three common switches found in microwave:

a) Primary switch
b) Monitor switch
c) Secondary switch.

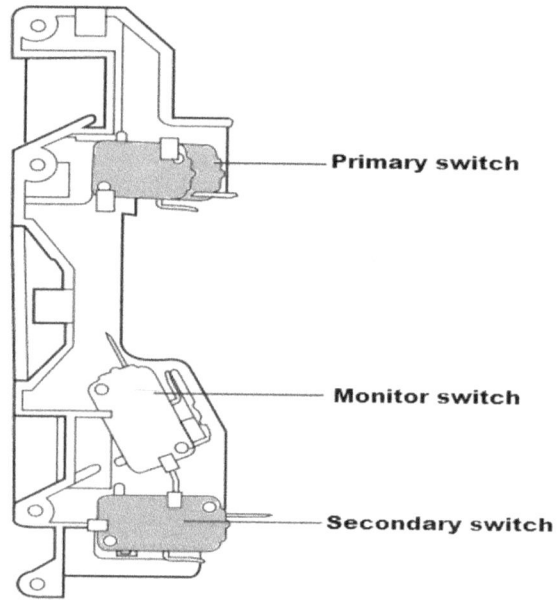

The monitor switch is usually found at the middle as shown above. As the name implies this switch is used to monitor the state of the microwave, if this switch is faulty the fuse will blow.

Testing microwave switches

When testing the switches, it is important to remove the wire lead connectors.
Before removing the wire connectors always be sure to take a photo of the connectors so that you will not connect them back wrongly.

Testing primary and secondary microwave switches with an ohmmeter.

Microwave door closed (switch wire lead removed)

When testing both the primary and secondary switch (top and bottom switch) you will get a beep if the switch is okay.
For the middle switch (monitor switch) the switch will read open circuit (high resistance) when the door is closed, the opposite of the other two switches.

Please note that if you test the monitor (middle) switch with the wire lead connected, it will read a low resistance due to the low resistance of the primary (side) of the mains transformer.

Microwave door Open (switch wire lead removed)

When you open the microwave door, the primary and secondary switch will read open circuit and the monitor(middle) switch will beep (short) if the switch is okay.

Summary.
How to test microwave switches

COMPONENTS	TEST		RESULTS	
	Test switch continuity as shown below		Door closed	Door open
SWITCHES (Wire leads removed)	Primary Switch	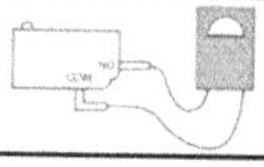	LOW	HIGH
	Monitor Switch	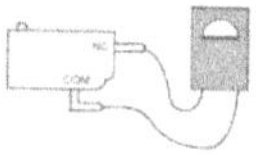	HIGH	LOW
	Secondary Switch	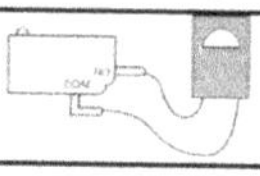	LOW	HIGH

High voltage circuit

All microwave ovens have a high voltage circuit and are composed of the following components.

1. High voltage transformer
2. High voltage capacitor
3. Magnetron
4. HV diode

If any of these components fail, then we will have no generation of microwave frequency used to heat food, and therefore you will have a microwave which seems to be heating but no heat produced.

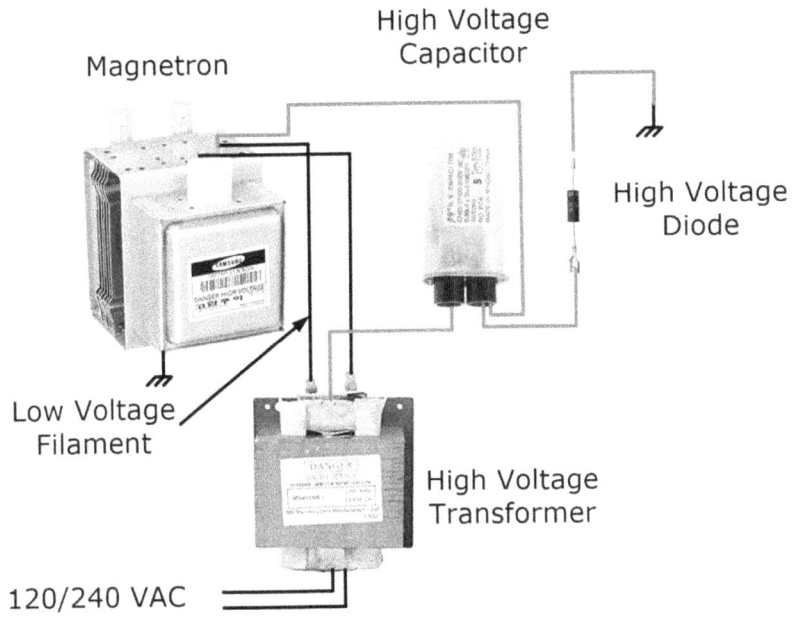

Microwave Oven pictorial high voltage circuit

High voltage transformer

When you lift the microwave oven you realize it is quite heavy, most of this weight can be attributed directly to the HV transformer.

As the name implies a transformer is used to transform or change something. They are usually used to step-up or step-down AC voltages, to change low voltage high current AC to high voltage low current AC.

The power input of a transformer is the same as power output (neglecting losses), this means you can only change the voltage and current but not the power since you cannot give what you do not have.

This means if I want the voltage to increase on the secondary side mean, then current will decrease on the secondary and the equation will balance, see the example below.

Power (primary)=V x I   equal   Power(secondary)=V x I

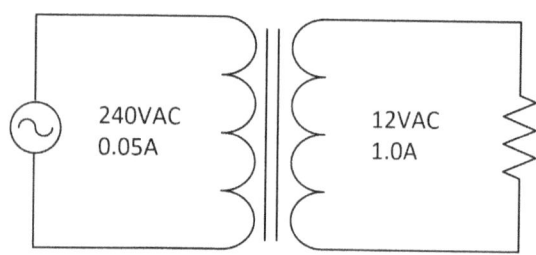

Primary P=V x I or 240 x 0.05 = 12Watts
Secondary P= V x I= 12x 1= 12 Watts.

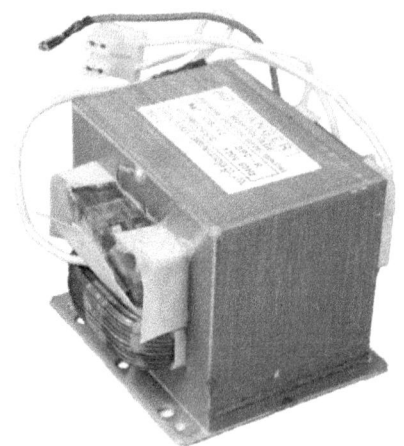

HV transformers, used in microwaves, have two secondary windings. One output has low voltage between 3.1 to 3.2 volts, while the high voltage output is around 5000-6000 volts.

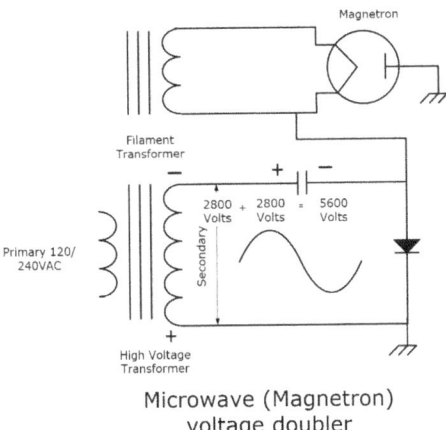

Microwave (Magnetron)
voltage doubler

The low voltage output is used to light the filament of the magnetron and the high voltage AC output from the transformer is changed to DC and doubled using a voltage doubling circuit (HV diode and HV capacitor) to power the magnetron.

Whenever working on microwaves it is important to be incredibly careful around these components. Follow all the safety measures when working around this circuit otherwise if you make a mistake, electrocution (which is fatal) will occur.

It is not advisable to turn on the microwave when the cover is removed. To do a live test, be sure to disconnect the magnetron connector so that the microwave oven does not work and expose you to the dangerous microwave frequency. Again, never test the output of HV transformer with a meter (NEVER!).

Testing HV transformers

You can use your analogue or digital meter to test if HV transformer is good or not. First you will need to identify different windings of this transformer which are labelled below.

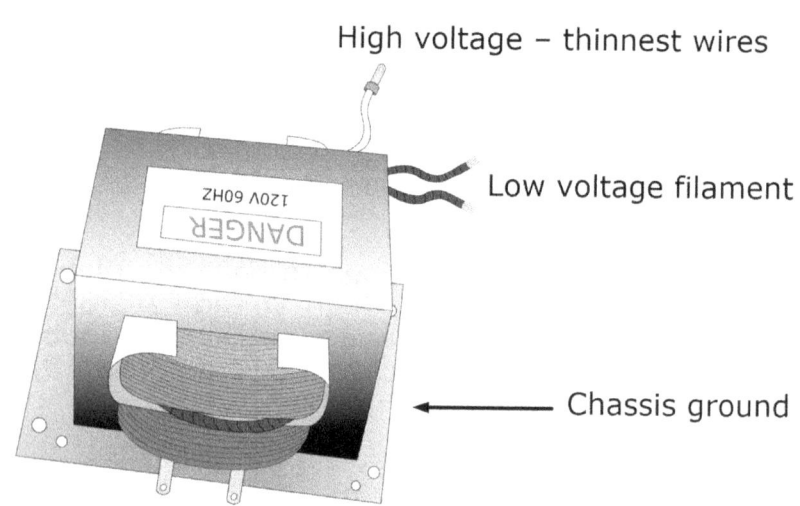

High voltage – thinnest wires

Low voltage filament

Chassis ground

Primary – thicker wire

Testing the primary coil with a digital meter I am getting 2.8 Ohms as you can see below. While testing any of the primary connector to the ground you will get an open circuit.

Microwave oven repair - made easy

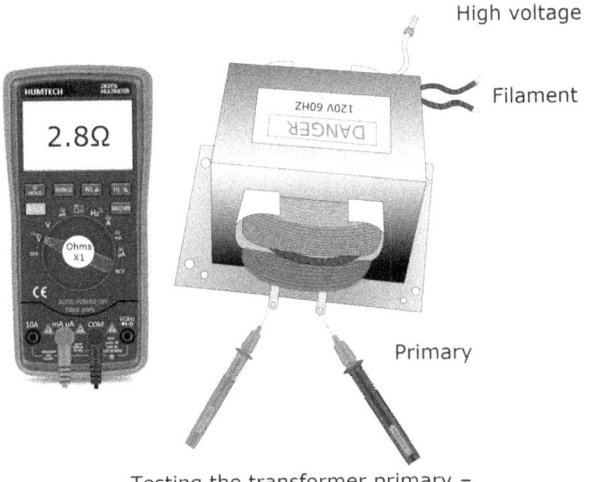

Testing the transformer primary – thicker conductor – less then 5Ω

Testing the filament winding with a digital meter I am getting 0.4 ohms. While testing between any of the filament connectors to the ground (bolt hole) you will get an open circuit if okay.

Testing the high voltage coil.

To test this coil, use the output, which usually has a fuse connected to it, as terminal one and for the other terminal use the transformer screw hole/screw.
If you use the transformer body, you will get an open circuit since this transformer is usually insulated, you can easily confirm this using continuity.
For this transformer I am getting 193.8Ω. This figure can vary for different models.

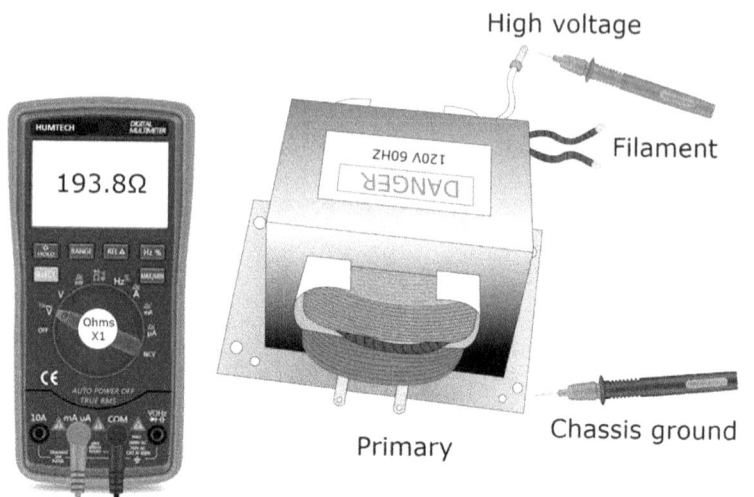

Testing the transformer high voltage winding – less then 35-200Ω

High voltage capacitor

An HV capacitor in a microwave oven is used as a voltage doubler, as the name implies, it doubles the output voltage. To achieve this an HV diode is used.

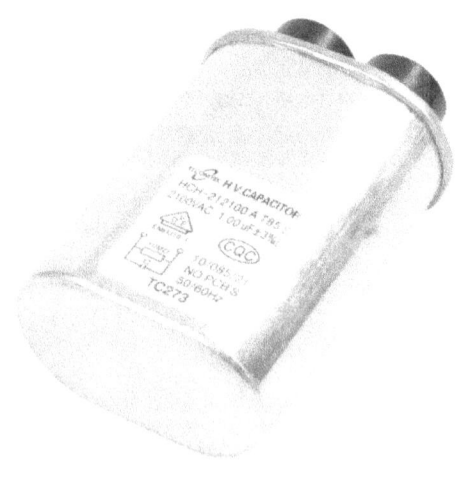

Testing an HV capacitor is easy, you can use a multi-component tester if you have one or use your analogue meter.

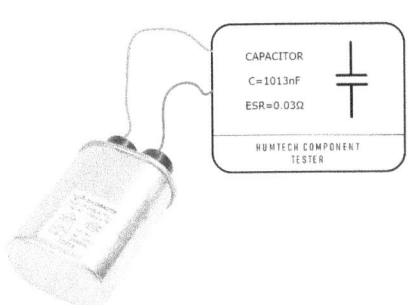

Testing the high voltage capacitor

For the analogue meter, set your meter to the highest range (10KΩ) and test across the two pins of the capacitor. If the capacitor is good, expect a small deflection and the meter to return to an open circuit.
To do this test be sure the HV capacitor is discharged.
To discharge the capacitor, short the pins of the capacitor to the chassis with a well-insulated screwdriver.

Magnetron

The magnetron is the heart of the microwave generating microwave energy used to heat the food.

Testing if good or bad

The first test for a magnetron is physical observation for any broken part. (I remember one time I struggled with repairing an LG microwave for a whole week which had "no heating symptoms" but everything else looked okay and I decided to change the magnetron directly. When I removed the magnetron, I noted the magnet part was broken into two pieces causing it not to heat food).

Next, use a digital meter, or an analogue meter, set to the lowest resistance settings.

When you test across the two filament terminals expect a low resistance of less than 1 ohm. Then set the meter to the highest resistance and test across the filament terminal and the chassis ground of the magnetron and expect an open circuit if it is good.

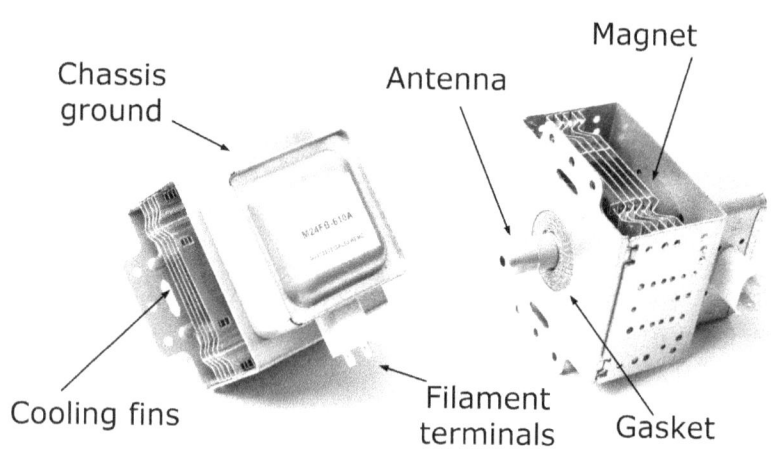

NB: the best test for a magnetron is direct replacement from my experience, especially if everything is found to be working okay and still no heating effect, just replace the magnetron and 90% percent of the time the heating problem will be rectified.

High voltage diode

Diodes are a one-way valve; this means they allow current to flow in one direction only.

Diodes used in microwaves are ultra-fast recovery rectifier diodes and therefore you cannot use any diode. You can salvage one from a junk microwave oven from the same circuit if not able to get one from the shop (market).

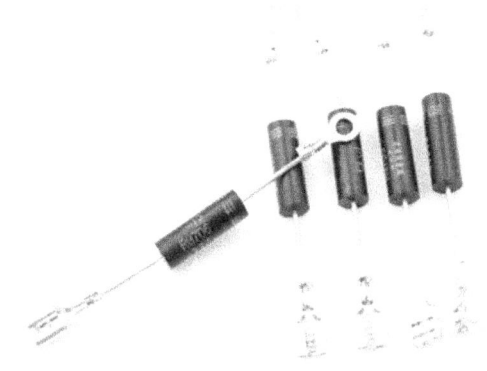

The big size of this diode means it can handle high power or current.

Testing HV diode with analogue meter is good enough to know if good or bad.

Please note. Use a high resistance setting of 10KΩ as a digital meter in diode mode will not work, neither will an analogue meter on a low resistance setting.

If a diode is good expect a small deflection in one direction and when you reverse the probes the needle will not move at all.

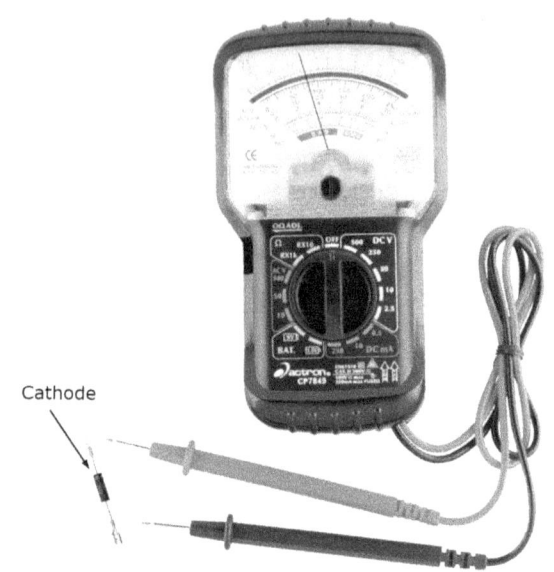

Small deflection as seen above.

Note: You cannot use a digital meter in this test as it will indicate the diode is open. If the analogue meter deflects either way, it means the diode is shorted.

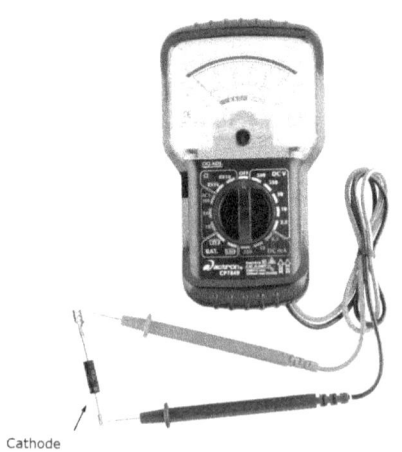

No deflection after reversing the probes.

Microwave oven repair - made easy

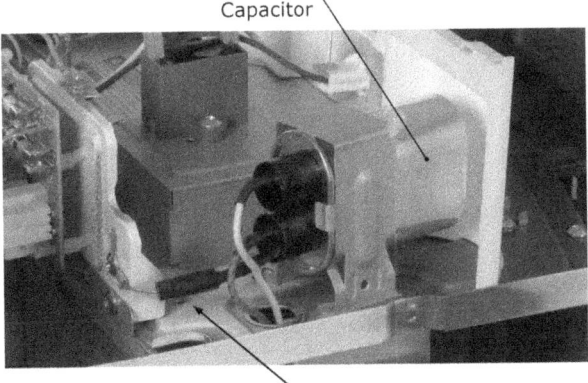

High Voltage Capacitor

High Voltage Diode

HV diode in a circuit - to do the test make sure the HV capacitor is discharged. Also disconnect one pin of the diode for correct results.

Some microwaves have a diode across the two pins of the HV Capacitor called ASYMMETRIC RECTIFIER. This diode prevents current flow in both directions.

Asymmetric Diode Symbol

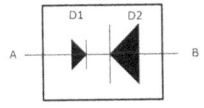

To test this diode set you analogue meter to the highest setting (10KΩ) and test between pin A and B. If the diode is good you should get an open circuit both ways.

If an asymmetric rectifier is shorted, (a reading in either direction) then the asymmetric rectifier is faulty and must be replaced with a high voltage asymmetric rectifier.

If the asymmetric rectifier is shorted, this could point to the magnetron, HV rectifier or HV Transformer being shorted.

FUSE

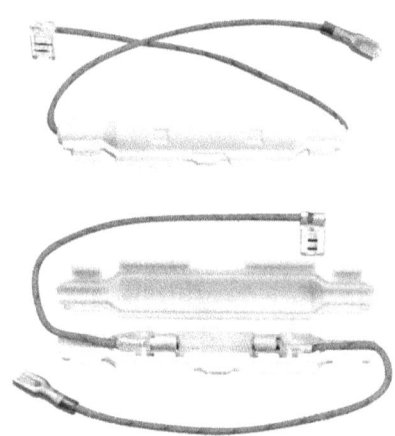

High voltage inline fuse

Microwave ovens' high voltage circuit have a high voltage fuse (see below diagram). The fuse is rated 5Kv by 0.7A and therefore it is highly covered by insulating material.

Whenever you find this fuse is blown, it could point to shorted HV circuit components. Before replacing the fuse, it is important to find out why the fuse opened since fuses are safety components meant to open (die) if something goes wrong (short) within the monitored component.

Also never replace a blown fuse with a jumper wire (thick) like I have seen done by other repairers.

Thermal cut-out

A thermal cutoff is an electrical safety device (either a thermal fuse or thermal switch) that interrupts electric current when heated to a specific temperature.

These devices may be for one-time use (a thermal fuse) or may be reset manually or automatically (a thermal switch).

If the temperature within the microwave exceeds certain temperatures, this component will open and cut off power supply to components of the microwave oven.

This rise in temperature usually occurs when the fan fails, or vents are blocked.

This explains why you should never put something on top of the microwave to avoid blocking these vents.

Here we have different thermal cutoff to monitor different circuits.

Usually, we have one which monitor the temperature of the magnetron and usually open if magnetron has overheated.

Another thermal cutoff monitors the grill temperature and usually opens if the cavity has overheated which could be caused by no load operation.

Testing if good or bad

Just like an ordinary fuse, continuity means it is good.

Turntable motor

This motor is used to turn the glass tray inside the microwave for uniform heating of food.

The motor is usually located at the bottom of the microwave. When this motor fails, it can cause a microwave to produce a grinding noise.

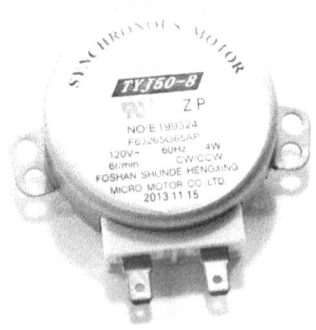

Testing if good or bad

Use an analogue or digital meter set to 2KΩ. If the motor is okay, it should measure less the 500Ω.
When testing be sure to remove the connectors otherwise you will get false readings.

Resistance should be below 500Ω

Fan motor

This fan is used to cool the magnetron when the microwave is working.

Testing:

Look for any sign of stress, worn out parts, blackened colour, especially the coil and damaged blades.
Testing the 3 terminals (any set) of the fan motor connectors by analogue or digital meter set to 2KΩ. Expect to get less than 500Ω if okay.
If you hear an unusual noise from the fan motor, it

could be the blades coming into contact with other objects.

Roller guide

Used to turn the glass tray inside the microwave cavity.

Testing:

Visually inspect for any damage, wear or cracked parts. When roller guide has a problem like a broken wheel, or not seated well, this can cause the microwave to produce some noise.

Microwave oven repair - made easy

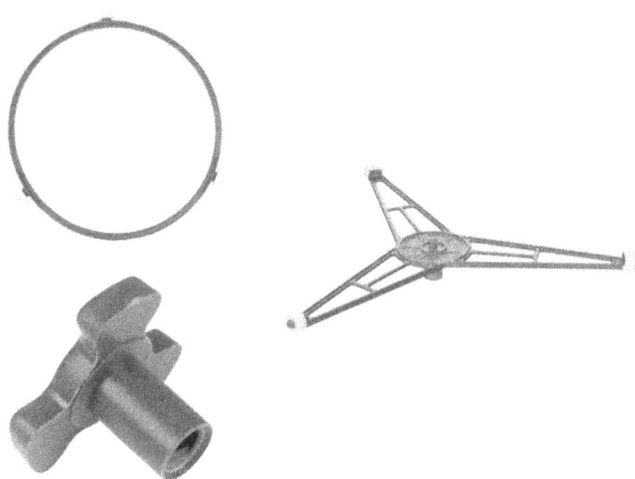

Summary:

Causes of noisy microwave includes:

- Turn table motor
- Magnetron
- Roller guide
- HV diode
- Fan motor

No heating problems

Causes of microwave "No heating symptoms".

Sometimes, during your repair work, you will get a microwave which to a layman appears to be working normally.

If the microwave fan, bulb and the front keyboard

are working, you can set the cooking time to say 10 seconds for example and press the start button. Then at the end of the 10 seconds you open the door only to find the food has not heated.

This is a common problem, and the causes can differ. From my experience, the Magnetron is most often the major cause for this, especially if you find the fuses are all intact.

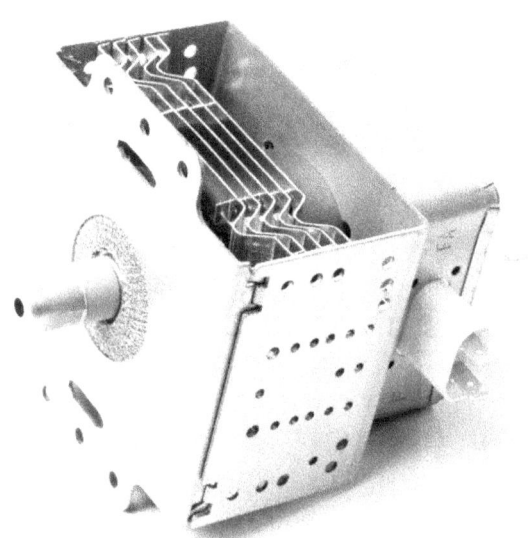

If after testing, you find it is okay but still not heating replace the part directly. Other causes of "no heating symptoms" in microwaves includes, HV diode, door switch, HV capacitor and HV transformer.

Microwave leakage test

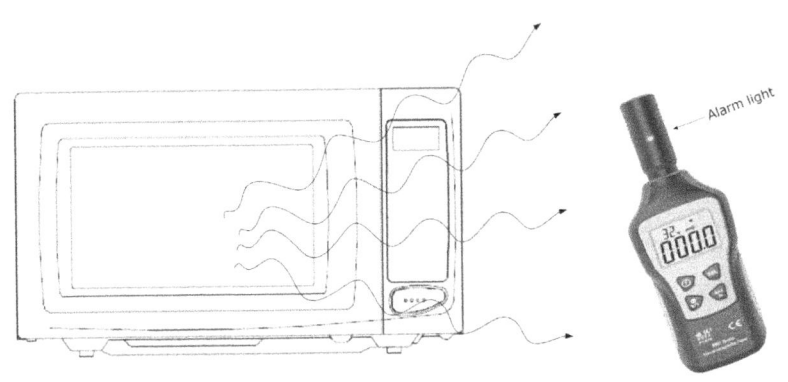

Electromagnetic Radiation in all directions

Move detector all around the microwave oven.

As a technician it is important to think about your safety and the safety of the user.

Microwave ovens operate at high frequencies which are harmful when exposed to human beings.

Therefore, it is important to test microwave radiation leakage to all microwave ovens brought to your shop even before opening and after the repair is complete before returning it back to the owner.
You can easily test for microwave radiation leakage using electromagnetic field radiation detector (E-Radiation tester).

There are a variety available on the market and therefore you can order depending on your budget.
For the model above the alarm light changes depending on amount of radiation leakage.
Green light for Low, Orange for Medium and Red for High.

To do the test, place the load in the oven, preferably a glass of water.

Close the door and turn the oven ON with a timer set to 2 minutes (if the water boils before the test is complete) you can replace the water and complete the test.

To test: Hover the tester antenna around the microwave body slowly, about 5cm from the surface of the microwave.

Pay close attention to vents, doors and door seals. If you detect any leakage around the door seals, lookout for worn out or loose door hinges or food particles built up around the door seals.

Measurement of Microwave output power

Just as we have microwave ovens in different sizes, colour and shapes, their cooking power also can vary from oven to oven.

Most microwave ovens have a power (wattage) range from 600 to 1200 watt.

The higher the wattage the higher the power.
To find out the output power of your microwave, fill a glass container with 1 litre of room temperature water (approx. 20°C).

Set the microwave on high power and note how long it will take the water to boil.

(a) 1-1/2 mins. 1200 Watts
(b) 2 mins. 1000 Watts
(c) 2-1/2 mins. 800 Watts
(d) 3 mins. 600 Watts

Using that that information you can adjust your microwave power level or cooking time to achieve more precise results.

Other books by this Author

Foundations Electricity & Electronics.
Available on Amazon paperback and Kindle.

ASIN : B08K3YHXS3

This book is for beginner studies in Electricity and Electronics. You will read this book if you are seeking to advance your career in electricity and electronics or perhaps to pursue a hobby. The book does not rush the reader. There are many diagrams and photographs to help you visualise your way. You are taken on a journey starting as a beginner. You will enjoy the fundamentals first and then concepts built upon concepts will lead you to a sound knowledge of the subject.

www.ingramcontent.com/pod-product-compliance
Lightning Source LLC
Chambersburg PA
CBHW070845220526
45466CB00002B/888